Contents

Words that appear in **bold** are explained in the glossary.

The answers to the questions are on pages 20-21.

What are materials?

Materials are what things are made of. We use different materials to make different objects.

Metal, wood, paper, glass, plastic and **cloth** are some of the materials that we use every day.

4

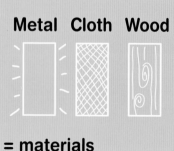

= materials

Now it's your turn...

Some objects are made of just one material.

Some objects are made of two or more different materials.

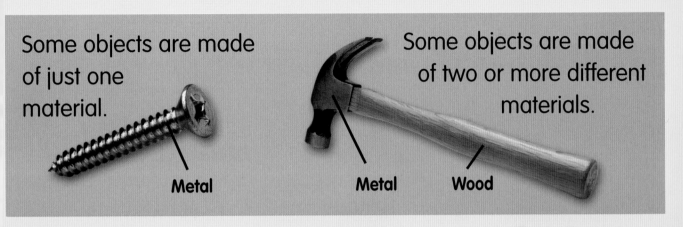

Metal

Metal   Wood

Metal, cloth and wood are materials.

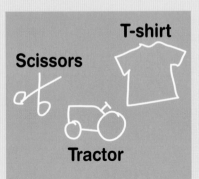

Scissors

T-shirt

Tractor

Look at the scissors, the T-shirt and the tractor. Are these objects or materials?

5

Stone

Stone is a hard, heavy **natural** material. It is the same material as the rocks that make mountains and cliffs.

Stone is dug out of the ground and from the sides of cliffs and mountains using huge diggers.

Stone lasts a long time. Many buildings are made from stone.

Stone = heavy and long-lasting

Now it's your turn...

Stone does not bend, but it can be cut into shapes or broken into pieces.

These huge stone faces have been cut into the side of a mountain in America.

Mud = soft

Stone

If you step in a patch of soft mud you will leave a footprint. What happens if you step on a block of stone?

7

Metal

Metal is a material made from some kinds of stone. The stone is smashed into powder. Then it is heated so the metal comes out of the stone.

Stone that contains metal is called ore. This is iron ore.

Metals are hard like stone, but they will bend.

Metal is a good material for making things like bendy fence wire.

Metal = hard and bendy

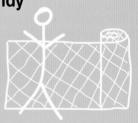

Metals are strong and difficult to break. They usually look shiny.

Look at all these different objects.

What will be left if you take all the metal objects away?

9

Wood

Wood is a natural material. It comes from trees that have been cut down.

Inside a tree trunk there are rings. A new ring grows each year.

If you count the rings, you can find out the age of the tree.

The rings make a pattern in wood that is called grain.

Metal saw = can cut wood

Now it's your turn...

Wood is a hard material, but it is not as hard as stone or metal. Wood can easily be cut and shaped with metal tools.

Wooden violin

Wooden chess game

Wooden saw

You can cut wood with a metal saw.
Can you cut metal with a wooden saw?

11

Paper

Paper is a soft, **artificial** material that is made from wood.

Pulp

The wood is chopped into tiny pieces and soaked in water to make a mushy mixture called **pulp**. The pulp is then squeezed out into thin sheets and dried.

Sheets of paper are light in weight and are very easy to bend and fold.

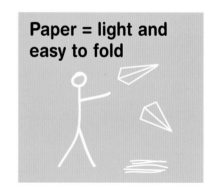

Paper = light and easy to fold

Now it's your turn...

You can make a chair from pieces of wood.

Paper chair

This chair has been made from pieces of paper. What do you think will happen if you sit on it?

13

Glass

Glass is a hard, artificial material that is made from sand. The sand is heated until it is very, very hot. Then it melts and turns into glass.

Hot glass is soft. It can be made into different shapes before it cools down and becomes hard.

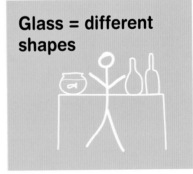

Glass = different shapes

Now it's your turn...

Glass is **transparent**. This means that light passes through it.

Glass windows let light in and allow us to see out.

You cannot see the drink inside a metal can. Can you see inside a glass bottle?

Plastic

Plastic is an artificial material that is made from **crude oil**. The oil is pumped out of the ground and sent to a **refinery**.

This machine is pumping oil from under the ground.

At the refinery, the oil is heated to produce plastic and other useful substances such as petrol.

Plastic = any shape or colour

16

Plastic can be made in any shape and colour.

Check it out
Now it's your turn...

Plastic can be bendy like a drinking straw, hard like a ball or soft like the trainers and rucksack.

Plastic can also be transparent. Which plastic objects in the picture can you see through?

Cloth

Cloth is a soft, artificial material made from **fibres**. Some cloth is made from plastic fibres, and some is made from natural fibres.

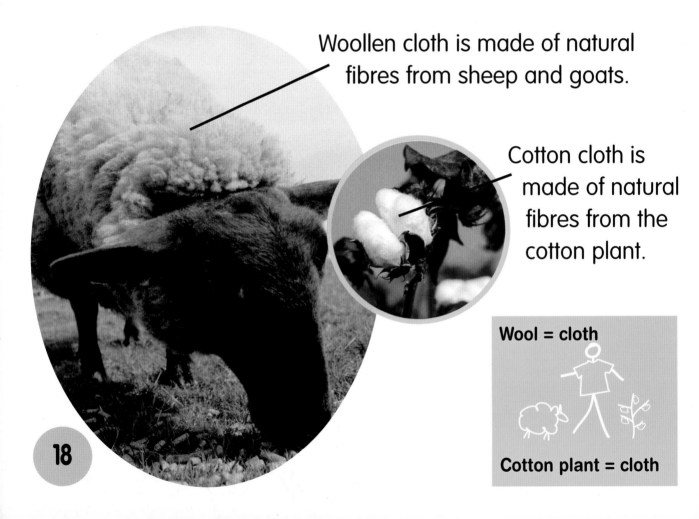

Woollen cloth is made of natural fibres from sheep and goats.

Cotton cloth is made of natural fibres from the cotton plant.

Wool = cloth

Cotton plant = cloth

The fibres are spun into a continuous **yarn**.
Then the yarn is woven or knitted into cloth.

The cloth becomes many things such as T-shirts, jeans and jumpers.

If you fold a sheet of paper in half it will stand on end. Can you do this with a piece of cloth?

19

Answers

Page 5

They are all objects made from materials. The scissors are an object made of metal, the T-shirt is an object made of cloth and the tractor is an object made of wood.

Scissors, T-shirt and tractor = objects

Page 7

Nothing will happen! If you step on a block of stone you will not leave a footprint because the surface of stone is much too hard.

Stone = hard

Page 9

There will be nothing left because they are all made from metal. The spanner, kettle, spoon and fork are made from steel. The trumpet is made from brass. The necklace is made from gold, and the can is made from aluminium.

Spanner, kettle, spoon, fork, trumpet, necklace and can = metal

20

Page 11

No you cannot cut metal with a wooden saw because metal is a harder material than wood.

Wooden saw = cannot cut metal

Page 13

It will collapse! Pieces of paper will fold and crumple if you put weight on them, so they are not strong enough to make a chair from.

Paper chair = not strong

Page 15

Glass is a transparent material so you can see the drink inside a glass bottle.

Glass bottle = see-through

Page 17

You can see through the bottle, the sandwich packet and the sunglasses.

Time for a snack!

Some plastic = see-through

Page 19

You cannot make a folded piece of cloth stand on end. Cloth is so soft, it collapses under its own weight. Most paper is soft, but paper can support its own weight when it is folded.

Cloth = soft

Glossary

artificial Anything that is made by people.

cloth A kind of material that is made from fibres joined together by weaving or knitting.

crude oil Oil that has been pumped from under the ground. It is natural and has not had anything done to it by people.

fibres Long, thin pieces of a material, such as plastic or wool.

glass A transparent material made from sand.

material Any substance that things are made from.

metal A hard, strong material, such as gold or iron.

natural Anything that is made entirely by nature.

paper A lightweight material made from wood.

plastic An artificial material made from crude oil. It can be made into almost anything.

pulp Something that is wet and mushy.

refinery A place where crude oil is taken to be made into other things.

stone A hard, heavy, natural material that can be dug from cliffs or mountains.

transparent See-through – allows light to pass through.

wood A natural material that comes from trees.

yarn A long thread made of fibres twisted together.

23

Index

Picture credits
t=top, b=bottom, c=centre, l=left, r=right, OFC=outside front cover
Corbis: 11, 12, 17 (dog), 18. Photodisc: OFC, 4, 5, 7, 10, 16, 17. Powerstock: OFC, 1, 3, 4, 8b, 13, 14, 15, 19.

Every effort has been made to trace the copyright holders, and we apologise in advance for any unintentional omissions. We would be pleased to insert the appropriate acknowledgements in any subsequent edition of this publication.

24